MIX
Papier aus verantwortungsvollen Quellen
Paper from responsible sources
FSC® C105338

Osama Mohammed Elmardi

Dynamic Relaxation Method

Theoretical Analysis, Solved Examples and Computer Programming

Anchor Academic
Publishing

Mohammed Elmardi, Osama: Dynamic Relaxation Method. Theoretical Analysis, Solved Examples and Computer Programming, Hamburg, Anchor Academic Publishing 2016

Buch-ISBN: 978-3-96067-084-1
PDF-eBook-ISBN: 978-3-96067-584-6
Druck/Herstellung: Anchor Academic Publishing, Hamburg, 2016

Bibliografische Information der Deutschen Nationalbibliothek:
Die Deutsche Nationalbibliothek verzeichnet diese Publikation in der Deutschen Nationalbibliografie; detaillierte bibliografische Daten sind im Internet über http://dnb.d-nb.de abrufbar.

Bibliographical Information of the German National Library:
The German National Library lists this publication in the German National Bibliography. Detailed bibliographic data can be found at: http://dnb.d-nb.de

All rights reserved. This publication may not be reproduced, stored in a retrieval system or transmitted, in any form or by any means, electronic, mechanical, photocopying, recording or otherwise, without the prior permission of the publishers.

Das Werk einschließlich aller seiner Teile ist urheberrechtlich geschützt. Jede Verwertung außerhalb der Grenzen des Urheberrechtsgesetzes ist ohne Zustimmung des Verlages unzulässig und strafbar. Dies gilt insbesondere für Vervielfältigungen, Übersetzungen, Mikroverfilmungen und die Einspeicherung und Bearbeitung in elektronischen Systemen.

Die Wiedergabe von Gebrauchsnamen, Handelsnamen, Warenbezeichnungen usw. in diesem Werk berechtigt auch ohne besondere Kennzeichnung nicht zu der Annahme, dass solche Namen im Sinne der Warenzeichen- und Markenschutz-Gesetzgebung als frei zu betrachten wären und daher von jedermann benutzt werden dürften.

Die Informationen in diesem Werk wurden mit Sorgfalt erarbeitet. Dennoch können Fehler nicht vollständig ausgeschlossen werden und die Diplomica Verlag GmbH, die Autoren oder Übersetzer übernehmen keine juristische Verantwortung oder irgendeine Haftung für evtl. verbliebene fehlerhafte Angaben und deren Folgen.

Alle Rechte vorbehalten

© Anchor Academic Publishing, Imprint der Diplomica Verlag GmbH
Hermannstal 119k, 22119 Hamburg
http://www.diplomica-verlag.de, Hamburg 2016
Printed in Germany

Dedication

In the name of Allah, the merciful, the compassionate

All praise is due to Allah and blessings and peace is upon his messenger and servant, **Mohammed**, and upon his family and companions and whoever follows his guidance until the day of resurrection.

To the memory of my mother **Khadra Dirar Taha**, my father **Mohammed Elmardi Suleiman**, and my dear aunt **Zaafaran Dirar Taha** who they taught me the greatest value of hard work and encouraged me in all my endeavours.

To my first wife **Nawal Abbas** and my beautiful three daughters **Roa, Rawan** and **Aya** whose love, patience and silence are my shelter whenever it gets hard.

To my second wife **Limya Abdullah** whose love and supplication to Allah were and will always be the momentum that boosts me through the thorny road of research.

To professor **Dr. Mahmoud Yassin Osman** for reviewing and modifying the manuscript before printing process.

This book is dedicated mainly to undergraduate and postgraduate students, especially mechanical and civil engineering students plus mathematicians and mathematics students where most of the applications are of mathematical nature.

To Mr. **Osama Mahmoud** of Daniya Center for printing services whose patience in editing and re – editing the manuscript of this book was the momentum that pushed me in completing successfully the present book.

To my homeland, Sudan, hoping to contribute in its development and superiority.

Finally, may Allah accepts this humble work and i hope that it will be beneficial to its readers

Acknowledgement

I am grateful and deeply indebted to Professor **Mahmoud Yassin Osman** for valuable opinions, consultation and constructive criticism, for without which this work would not have been accomplished.

I am also indebted to published texts in the analysis of dynamic relaxation method which have been contributed to the author's thinking. Members of Mechanical Engineering Department at Faculty of Engineering and Technology, Nile Valley University, Atbara – Sudan, and Sudan University of Science & Technology, Khartoum – Sudan have served to sharpen and refine the treatment of my topics. The author is extremely grateful to them for constructive criticisms and valuable proposals.

I express my profound gratitude to Mr. **Osama Mahmoud** of Daniya Center for computer and printing services, Atbara, who spent several hours in editing, re – editing and correcting the present manuscript.

Special appreciation is due to the British Council's Library for its quick response in ordering the requested bibliography, books, reviews and papers.

Preface

Chapter one includes introduction to dynamic relaxation method (DR) which is combined with finite differences method (FD) for the sake of solving ordinary and partial differential equations, as a single equation or as a group of differential equations. In this chapter the dynamic relaxation equations are transformed to artificial dynamic space by adding damping and inertia effects. These are then expressed in finite difference form and the solution is obtained through iterations.

In chapter two the procedural steps in solving differential equations using DR method were applied to the system of differential equations (i.e. ordinary and/ or partial differential equations). The DR program performs the following operations: Reads data file; computes fictitious densities; computes velocities and displacements; checks stability of numerical computations; checks convergence of solution; and checks wrong convergence. At the end of this chapter the dynamic relaxation (DR) numerical method coupled with the finite differences discretization technique is used to solve nonlinear ordinary and partial differential equations. Subsequently, a FORTRAN program is developed to generate the numerical results as analytical and/ or exact solutions.

The book is suitable as a textbook for a first course on dynamic relaxation technique in civil and mechanical engineering curricula. It can be used as a reference by engineers and scientists working in industry and academic institutions.

Author

Assistant Professor

Osama Mohammed Elmardi Suleiman

Contents

Dedication	i
Acknowledgement	ii
Preface	iii
Contents	iv
Symbols and Abbreviations	v

Chapter One: Introduction

1.1 General Introduction	1
1.2 Formulation of Dynamic Relaxation Equations	1
1.3 Finite Difference Approximation	4
1.3.1 Ordinary Differential Equations	4
1.3.2 Partial Differential Equations	7

Chapter Two: Procedural Steps in Solving Differential Equations Using Dynamic Relaxation (DR) Method

2.1 The Dynamic Relaxation (DR) Program	11
2.1.1 Numerical Instability	11
2.1.2 Convergence of DR Solution	11
2.1.3 Convergence to an Invalid Solution	12
2.1.4 Time Increment	12
2.1.5 Damping Coefficient	12
2.2 Solved Examples	12
2.2.1 Solution of Ordinary Differential Equations	13
2.2.2 Solution of Partial Differential Equations	17
Bibliography	31

Symbols and Abbreviations

$KS = K^*$

$NMAX$ = maximum number of iterations

EQ = equal to

Dx = discretization of solution in the x – direction

GT = greater than

LT = less than

$\dfrac{\partial u}{\partial t}$ = velocity

$\dfrac{\partial^2 u}{\partial t^2}$ = acceleration

Δt = time increment

ρ = inertia effect

k = damping effect

Chapter One

Introduction

1.1 General Introduction:

Dynamic Relaxation method (DR) Coupled with Finite Differences method (FD) is used for solving ordinary and partial differential equations as a single equation or as a group of differential equations. To apply dynamic relaxation software technique, the differential equations are transformed into dynamic equations by adding damping and inertia elements. These in turn are expressed in finite differences form, and the solution is obtained by an iterative procedure as is explained in the following paragraphs:

The differential equation is referred to in the following as:

$$f = 0 \qquad (1.1)$$

Where, f = 0, may be an ordinary differential equation as follows:

$$P(x)\frac{d^2u}{dx^2} + Q(x)\frac{du}{dx} + R(s)u = 0$$

Or a partial differential equation as stated below:

$$P(x)\frac{\partial^2 u}{\partial x^2} + Q(x)\frac{\partial^2 u}{\partial y^2} + R(x,y)u = 0$$

1.2 Formulation of Dynamic Relaxation Equations:

The dynamic relaxation method (DR) formula begins with the dynamic equation which may be written as:

$$f = \rho\frac{\partial^2 u}{\partial t^2} + k\frac{\partial u}{\partial t} \qquad (1.2)$$

In this procedure the statical system i.e. equation (1.1) is transferred to an artificial dynamic space by adding fictitious inertia and damping forces as in equation (1.2).

The DR method was first proposed in 1960s; refer to Rushton [1], Cassel and Hobbs [2], and Day [3]. In this method, the equations of equilibrium are converted to dynamic equations by adding damping and inertia terms, these are then expressed in finite difference form and solution is obtained through iterations. The optimum damping coefficient and time increment used to stabilize the solution depend on a number of factors including the stiffness matrix of the structure, the applied load, the boundary conditions and the size of the mesh used, etc.

In order to analyze various complicated problems in engineering, many kinds of efficient numerical methods such as finite difference method, finite element method and the weighted residual method have been developed. However, the accompanying problem is that large computers are needed to solve the related large scale equations. Sometimes, the equations are so large that one can only obtain rough results. This is especially conspicuous in solving non – linear problems. In addition, numerical instability during iteration is often involved.

In the traditional methods of solving equations from static equilibrium problems, it is considered that internal forces exist initially in the structures. In so doing, one assumes that the external forces were exerted very slowly so that the dynamic process of the structures could be neglected. In fact, as has been pointed out by Rayleigh [4], static solution of a mechanics system can be referred to as the steady state part of the transient response of the system to step loading. This approach was successfully applied to solving linear problems by Otter [5] and Day [3] in dependently in 1965, and was named the dynamic relaxation (DR) method.

Nowadays, researchers are attracted by the efficiency of solving non – linear problems with DR. The applications of DR to various problems indicate that the method has the following distinctive features { see, for example, [6] – [9] }.

Numerical techniques other than the dynamic relaxation (DR) method include finite element method (FEM), which is widely used in most of the theoretical analyses of today's research. In a comparison between the dynamic relaxation method and the finite element method, Aalami [10] found that the computer time required for finite element method is eight times greater than that for the dynamic relaxation analysis, whereas storage capacity for finite element analysis is ten times or more than that for DR analysis. This fact is supported by Putcha and Reddy [11], and Turvey and Osman { [12] – [14] }, who they noted that some of the finite element formulations require large storage capacity and computer time. However, if the analysis requires less computations and computer time, then, the dynamic relaxation is considered more efficient than the finite element method. In another comparison Aalami [10] found that the difference in accuracy between one version of finite element and another may reach a value of 10% or more, whereas a comparison between one version of finite element method and DR showed a difference of more than 15%. Therefore, the dynamic relaxation method (DR) can be considered of acceptable accuracy.

The only apparent limitation of dynamic relaxation (DR) method is that it can only be applied to limited geometries. However, this limitation is irrelevant to square and rectangular plates and beams which are widely used in engineering applications.

The errors inherent in the dynamic relaxation (DR) technique $\{[15] - [23]\}$ include discretization error which is due to the replacement of a continuous function with a discrete function. Also, there is an additional error resulting from the non – exact solution of the discrete equations due to the variations of the velocities from the edges of the plate to the center. The usage of finer meshes reduces the discretization error, but increases the round – off error due to the large amount of computations involved.

For the sake of simplifying and explanation of the DR method, u in equation (1.2) is referred to as displacement, and hence the terms $\partial u / \partial t$ and $\partial^2 u / \partial t^2$ are the velocity and acceleration respectively. Accordingly the first and second terms on the right – hand side are the inertia and damping terms respectively. ρ and k are the inertia and damping coefficients respectively, and t is time.

If the velocities before and after the period Δt at an arbitrary node in the finite difference mesh are denoted by $\{\partial u/\partial t\}_{n-1}$ and $\{\partial u/\partial t\}_n$ respectively, then using finite differences in time, and specifying the value of the function at $(n - \frac{1}{2})$, it is possible to write equation (1.2) in the following form:

$$f_{n-\frac{1}{2}} = \frac{\rho}{\Delta t}\left[\left\{\frac{\partial u}{\partial t}\right\}_n - \left\{\frac{\partial u}{\partial t}\right\}_{n-1}\right] + k\left\{\frac{\partial u}{\partial t}\right\}_{n-\frac{1}{2}} \qquad (1.3)$$

Now $\left\{\frac{\partial u}{\partial t}\right\}_{n-\frac{1}{2}}$, which is the velocity at the middle of the time increment, can be approximated by the mean velocities before and after the time increment, Δt, which is expressed as follows:

$$\left\{\frac{\partial u}{\partial t}\right\}_{n-\frac{1}{2}} = \frac{1}{2}\left[\left\{\frac{\partial u}{\partial t}\right\}_n - \left\{\frac{\partial u}{\partial t}\right\}_{n-1}\right]$$

Hence, equation (1.3) can be expressed in the following form as:

$$f_{n-\frac{1}{2}} = \frac{\rho}{\Delta t}\left[\left\{\frac{\partial u}{\partial t}\right\}_n - \left\{\frac{\partial u}{\partial t}\right\}_{n-1}\right] + \frac{k}{2}\left[\left\{\frac{\partial u}{\partial t}\right\}_n - \left\{\frac{\partial u}{\partial t}\right\}_{n-1}\right] \quad (1.4)$$

Equation (1.4) can then be arranged to give the velocity after the time interval, Δt:

$$\left\{\frac{\partial u}{\partial t}\right\}_n = (1+k^*)^{-1}\left[\frac{\Delta t}{e}f_{n-\frac{1}{2}} + (1-k^*)\left\{\frac{\partial u}{\partial t}\right\}_{n-1}\right] \quad (1.5)$$

Where:

$$k^* = \frac{k\Delta t}{2\rho}$$

The displacements at the middle of the next time increment can be determined by integrating the velocity, so that:

$$u_{n+\frac{1}{2}} = u_{n-\frac{1}{2}} + \left\{\frac{\partial u}{\partial t}\right\}_n \Delta t \quad (1.6)$$

The iterative procedure begins at time $t = 0$ with all initial values of the velocities and displacements equal to zero or any other suitable values. In the first iteration, the velocities are obtained from equation (1.5) and the displacements from equation (1.6). The boundary conditions are then applied. Subsequent iterations follow the same steps until the desired accuracy is achieved.

1.3 Finite Difference Approximation:

1.3.1 Ordinary Differential Equations:

The values of the interpolating function $u(x)$ in the vicinity of the node i in a non – uniform or graded mesh shown in figure (1.1) below can be expressed as follows using Taylor's series:

$$u(i+1) = u(i) + P_{i+1}\Delta x\, u'(i) + \frac{P_{i+1}^2 \Delta x^2}{2!}u''(i)$$

$$+\frac{P_{i+1}^3 \Delta x^3}{3!}u'''(i) + \frac{P_{i+1}^4 \Delta x^4}{4!}u''''(i) + \cdots \qquad (1.7)$$

$$u(i-1) = u(i) - P_i \Delta x \, u'(i) + \frac{P_i^2 \Delta x^2}{2!}u''(i)$$

$$-\frac{P_i^3 \Delta x^3}{3!}u'''(i) + \frac{P_i^4 \Delta x^4}{4!}u''''(i) + \cdots \qquad (1.8)$$

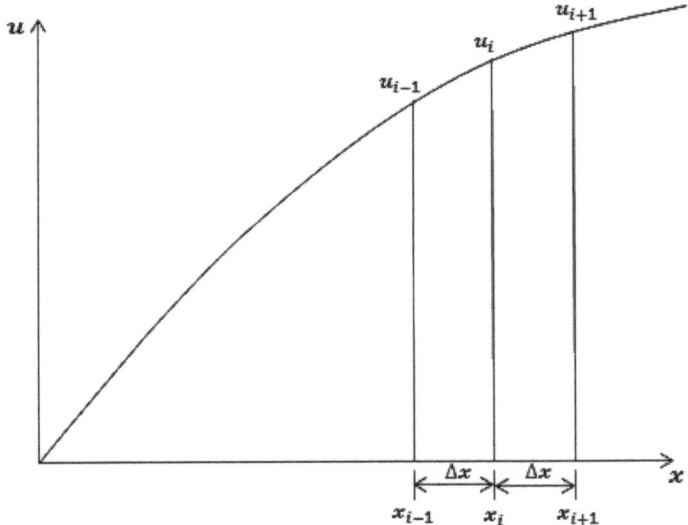

Figure (1.1) Non – uniform or graded mesh

Where $u'(i)$, $u''(i)$, $u'''(i)$, and $u''''(i)$ are the first, second, third, and fourth derivatives of the function $u(x)$ at node i.

When multiplying equation (1.7) by P_i^2 and equation (1.8) by P_{i+1}^2, then subtract the latter from the former and rearrange the resulting expression to obtain the function at node i as follows:

$$u'(i) = \frac{1}{\Delta x}\left[\alpha_i^{(1)} u(i+1) + \alpha_i^{(2)} u(i) + \alpha_i^{(3)} u(i-1)\right] + \epsilon_1 \qquad (1.9)$$

Where:

$$\alpha_i^{(1)} = \frac{P_i}{P_{i+1}(P_i + P_{i+1})}$$

$$\alpha_i^{(2)} = -\frac{p_i - p_{i+1}}{P_i P_{i+1}}$$

$$\alpha_i^{(3)} = -\frac{P_{i+1}}{p_i(p_i + P_{i+1})}$$

$$\epsilon_1 = -p_i\, p_{i+1} \frac{\Delta x^2}{6} u i''''' + \cdots \qquad (1.10)$$

Multiply equation (1.7) by p_i and equation (1.8) by p_{i+1} and add them together to obtain the second derivative of the function $u(x)$ at node i as follows:

$$u''(i) = \frac{2}{\Delta x^2}\left[\alpha_i^{(4)} u(i+1) + \alpha_i^{(5)} u(i) + \alpha_i^{(6)} u(i-2)\right] + \epsilon_2 \qquad (1.11)$$

Where:

$$\alpha_i^{(4)} = \frac{1}{P_{i+1}(p_i + p_{i+1})}$$

$$\alpha_i^{(5)} = -\frac{1}{P_i P_{i+1}}$$

$$\alpha_i^{(6)} = \frac{1}{p_i(p_i + p_{i+1})}$$

$$\epsilon_2 = \frac{P_i - P_{i+1}}{3} \Delta x u''' i - \frac{p_i^2 - p_i p_{i-1} + p_{i+1}}{12} \Delta x^2 u'''' i + \cdots \qquad (1.12)$$

If the derivatives of $f(x)$ which are greater than the third are assumed negligible i.e. the actual function approximates a quadratic function then ϵ_1 and ϵ_2 represent the error in the approximation resulting from replacing the actual function by a quadratic function. The error in the first derivative of the function of equation (1.10) depends on the graded mesh and it is proportional to Δx^2. The error in the second derivative of the function, equation (1.12), is proportional to Δx^2 for a uniform mesh (i.e. $P_i = P_{i+1}$), and proportional to Δx for a graded mesh (i.e. $P_i \neq P_{i+1}$). That is to say the error associated with a graded mesh is greater than that of a uniform mesh with the same number of elements. However, a graded mesh is more flexible than a uniform mesh and it allows closer nodes to be employed in those regions where a higher degree of accuracy is required.

When the mesh is uniform $P_i = P_{i+1}$, and hence:

$$\alpha_i^{(1)} = \alpha_i^{(4)} = \frac{1}{2}$$

$$\alpha_i^{(1)} = 0$$

$$\alpha_i^{(1)} = \alpha_i^{(6)} = -\frac{1}{2}$$

$$\alpha_i^{(5)} = -1$$

And therefore, the first and second derivatives, with the error neglected, are as follows:

$$\frac{du}{dx}(i) = \frac{1}{2\Delta x}[u(i+1) - u(i-1)] \qquad (1.13)$$

$$\frac{d^2u}{dx^2}(i) = \frac{1}{\Delta x^2}[u(i+1) - 2u(i) + u(i-1)] \qquad (1.14)$$

The first derivative of the function with respect to x can be written also for a uniform mesh as follows with:

$$\frac{du}{dx}(i) = \frac{1}{\Delta x}[u(i) - u(i-1)]$$

or

$$\frac{du}{dx}(i) = \frac{1}{\Delta x}[u(i+1) - u(i)] \qquad (1.15)$$

1.3.2 Partial Differential Equations:

The first and second derivatives of a function $u(x, y)$ at an arbitrary node (i, j) shown in figure (1.2) below can be written as follows:

$$\frac{\partial u}{\partial x}(i,j) = \frac{1}{\Delta x}\left[\alpha_{ij}^{(1)}u(i+1,j) + \alpha_{ij}^{(2)}u(i,j) + \alpha_{ij}^{(3)}u(i-1,j)\right]$$

$$\frac{\partial^2 u}{\partial x^2}(i,j) = \frac{2}{\Delta x^2}\left[\alpha_{ij}^{(4)}u(i+1,j) + \alpha_{ij}^{(5)}u(i,j) + \alpha_{ij}^{(6)}u(i-1,j)\right]$$

$$\frac{\partial u}{\partial x \partial y}(i,j) = \frac{1}{\Delta x \Delta y}\left[\alpha_{ij}^{(1)}\beta_{ij}^{(1)}u(i+1,j+1) + \alpha_{ij}^{(1)}\beta_{ij}u(i+1,j-1)\right.$$

$$\left. + \alpha_{ij}^{(3)}\beta_{ij}^{(1)}u(i-1,j+1) + \alpha_{ij}^{(3)}\beta_{ij}^{(3)}u(i-1,j-1)\right]$$

$$\frac{\partial u}{\partial y}(i,j) = \frac{1}{\Delta y}\left[\beta_{ij}^{(1)}u(i,j+1) + \beta_{ij}^{(2)}u(i,j) + \beta_{ij}^{(3)}u(i,j-1)\right]$$

$$\frac{\partial^2 u}{\partial y^2} = \frac{2}{\Delta y^2}\left[\beta_{ij}^{(4)}u(i,j+1) + \beta_{ij}^{(5)}u(i,j) + \beta_{ij}^{(6)}u(i,j-1)\right]$$

Where:

$$\alpha_{ij}^{(1)} = \frac{P_i}{P_{i+1}(P_{i+1}+P_i)}$$

$$\alpha_{ij}^{(2)} = \frac{P_{i+1}-P_i}{P_i P_{i+1}}$$

$$\alpha_{ij}^{(3)} = -\frac{P_{i+1}}{P_i(P_i+P_{i+1})}$$

$$\alpha_{ij}^{(4)} = \frac{1}{P_i+P_{i+1}}$$

$$\alpha_{ij}^{(5)} = -\frac{1}{P_i P_{i+1}}$$

$$\alpha_{ij}^{(6)} = \frac{1}{P_i(p_i+P_{i+1})}$$

$$\beta_{ij}^{(1)} = \frac{r_i}{r_{i+1}(r_{i+1}+r_i)}$$

$$\beta_{ij}^{(2)} = \frac{r_{i+1}-r_i}{r_i r_{i+1}}$$

$$\beta_{ij}^{(3)} = -\frac{r_{i+1}}{r_i(r_i+r_{i+1})}$$

$$\beta_{ij}^{(4)} = \frac{1}{r_i+r_{i+1}}$$

$$\beta_{ij}^{(5)} = -\frac{1}{r_i r_{i+1}}$$

$$\beta_{ij}^{(6)} = \frac{1}{r_i(r_i+r_{i+1})}$$

Where P_i and P_{i+1} are the ratios of the dimensions of the elements on both sides of the node (i,j) to the average element length all measured in the x – direction. r_i and

r_{i+1} are the ratios of the dimensions of the elements on both sides of node (i,j) to the average element length all measured in the y – direction.

	$i-1, j+1$	$i, j+1$	$i+1, j+1$
$r_{j+1}\Delta y$	$i-1, j$	i, j	$i+1, j$
$r_j \Delta y$	$i-1, j-1$	$i, j-1$	$i+1, j-1$
	$P_i \Delta x$	$P_{i+1} \Delta x$	

Figure (1.2) the first and second derivatives of a two dimensional function

$$u(x, y)$$

When the mesh is uniform i.e. $p_i = P_{i+1}$ and $r_i = r_{i+1}$, we have:

$$\alpha_{ij}^{(1)} = \beta_{ij}^{(1)} = \frac{1}{2}$$

$$\alpha_{ij}^{(2)} = \beta_{ij}^{(2)} = 0$$

$$\alpha_{ij}^{(3)} = \beta_{ij}^{(3)} = -\frac{1}{2}$$

$$\alpha_{ij}^{(4)} = \beta_{ij}^{(4)} = \frac{1}{2}$$

$$\alpha_{ij}^{(5)} = \beta_{ij}^{(5)} = -1$$

$$\alpha_{ij}^{(6)} = \beta_{ij}^{(6)} = \frac{1}{2}$$

The first and second derivatives of $u(x, y)$ for a uniform mesh are:

$$\frac{\partial u}{\partial x}(i,j) = \frac{1}{2\Delta x}[u(i+1,j) - u(i-1,j)] \qquad (1.16)$$

$$\frac{\partial u}{\partial y}(i,j) = \frac{1}{2\Delta y}[u(i,j+1) - u(i,j-1)] \qquad (1.17)$$

$$\frac{\partial^2 u}{\partial x^2}(i,j) = \frac{1}{\Delta x^2}[u(i+1,j) - 2u(i,j) + u(i-1,j)] \qquad (1.18)$$

$$\frac{\partial^2 u}{\partial y^2}(i,j) = \frac{1}{\Delta y^2}[u(i,j+1) - 2u(i,j) + u(i,j-1)] \qquad (1.19)$$

$$\frac{\partial^2 u}{\partial x \partial y}(i,j) = \frac{1}{4\Delta x \Delta y}[u(i+1,j+1) - u(i+1,j-1)$$
$$-u(i-1,j+1) + u(i-1,j-1)] \tag{1.20}$$

Chapter Two
Procedural Steps in Solving Differential Equations Using DR Method

2.1 The Dynamic Relaxation (DR) Program:

The DR program performs the following operations:
1. Reads data file.
2. Computes fictitious densities.
3. Computes velocities and displacements.
4. Checks stability of numerical computations.
5. Checks convergence of solution.
6. Checks wrong convergence.

Refer to references {[24] – [30]} for more information about analysis of rectangular laminated plates in bending.

2.1.1 Numerical Instability:

In every iteration, the value of the function at the center of the solution domain or other suitable point is compared with two estimated reference values representing lower and upper bounds of the function at that point. If solution was failed such that the computed value of the function at the specified point did not fall within the prescribed range, the solution is deemed unstable, and therefore iterations are terminated. The damping coefficients are then reduced and the process of iteration is restarted once again. The iterations are repeated several times until stability is reached.

2.1.2 Convergence of DR Solution:

Convergence of the dynamic relaxation solution is checked at the end of each iteration by comparing the velocities over the domain with a prescribed value. The procedure is repeated until the solution is deemed converged and consequently the iterative process is terminated.

2.1.3 Convergence to an Invalid Solution:

Sometimes DR solution converges to incorrect answer. Check for invalid solution is carried out after the solution has satisfied the convergence criterion explained earlier. In the check procedure the profile of variable is compared with the anticipated profile over the domain. For instance, if the value of the function on the boundaries is zero, and it is known that the function increases from edge to center, then the solution should follows a similar profile. If the computed profile is different from that, the solution is deemed to be incorrect. When this happens, the solution can hardly be made to converge to the correct answer by altering the damping coefficients and time increment. One should take another look to the boundary conditions and correct them if they are wrong.

2.1.4 Time Increment:

Proper time increment is a very important factor for speeding convergence and controlling numerical computations. When time increment is too small, convergence becomes tediously slow; and if it is too large, the solution becomes unstable. Time increment must be less than 1, say, 0.8 .

2.1.5 Damping Coefficient:

The optimum damping coefficient is that which produces critical motion. When the damping coefficient or coefficients are large, the motion is over – damped and convergence becomes very slow. When the coefficients are small, the motion is under – damped and can cause numerical instability.

2.2 Solved Examples:

In the following examples, the dynamic relaxation (DR) numerical method combined with the finite differences discretization technique is used to solve nonlinear ordinary and partial differential equations. Subsequently a FORTRAN program is developed to generate the numerical results as analytical and/ or exact solutions.

2.2.1 Solution of Ordinary Differential Equation:
Example (2.1):

Solve the following ordinary differential equation using the dynamic relaxation (DR) method.

$$\frac{d^2w}{dx^2} + q = 0 \qquad (2.1)$$

Where $q = \pi^2 \sin(\pi x)$, and the end conditions are:

$$w(0) = w(1) = 0$$

Note that the exact solution is $w = \sin(\pi x)$.

Solution:

Write equation (2.1) in finite difference form as shown below:

$$f = \frac{1}{\Delta x^2}[w(i+1) - 2w(i) + w(i-1)] + q(i)$$

The velocities are:

$$w_t(i)_n = \frac{1}{1+k^*(i)}\left[1 - k^*(i) w_t(i)_{n-1} + \frac{f_{n-\frac{1}{2}} \Delta t}{\rho(i)}\right]$$

The values of the function are computed from:

$$w(i)_{n+\frac{1}{2}} = w(i)_{n-\frac{1}{2}} + w_t(i)_n \Delta t$$

Now if the region of the problem $\{0 - 1\}$ is divided into 10 elements, then the end conditions can be expressed as:

$$w(0) = w(10) = 0$$

However, in this case and due to symmetry of end conditions, the solution can be obtained over half the domain {i.e. 0 – 1/2}. The condition at the symmetry line defined by $i = 5$ is:

$$w(6) = w(4)$$

All initial values are set to zero and iterations are started. After each iteration the velocities are compared with a reference of very small value of about 10^{-6}. When all velocities are less than the prescribed value, the process is terminated. The process may be terminated of course when the maximum value of the function (i.e. at center) exceeds certain bounds which indicates that the solution is becoming unstable. These bounds are defined by the inequalities $0 \leq w \leq 5$. After the solution has converged, a further check is made to guarantee that the solution has converged correctly. To facilitate this use is made of the fact that function increases from end to center. In fact this profile is achieved by the converged solution and therefore the solution is considered to be correct.

The computer output is listed in table (2.1) below. the solution was converged in 136 iterations. Note the close comparison between the approximate and exact solutions.

Table (2.1) solution of example (1)

x	0.10	0.20	0.30	0.40	0.50
w (approximate solution)	0.3119	0.5932	0.8164	0.9598	1.0091
w (exact solution)	0.3091	0.5880	0.8092	0.9512	1.0000

The FORTRAN program entitled [Osama 1. FOR] is shown below which is used to solve an ordinary differential equation of example (2.1) using DR method.

Osama 1. FOR

```
C   This program solves an ordinary differential equation using DR
C   method
    Real K, KS
    Dimension X(0:10), Q(0:10), W(0:10), WT(0:10)
    Open (unit = 5, File = 'Osama 1. Dat', status = 'old')
    Open (unit = 6, File = 'Osama 1. Out', status = 'unknown')
    Read (5, *) K, DT, RHO, NMAX
8   Format (1x, 'x', 8x, 'w')
9   Format (F4.2, 5x, F6.4)
    Dx = 0.1
    PI = 3.1416
    KS = K * DT / (2.0 * RHO)
    DO 10 I = 0,10
```

  x (I) = Dx * I
  Q (I) = (PI ** 2) * sin (PI * x(I))
10 Continue
C Initial Conditions
  Do 20 I = 0,10
  WT (I) = 0.0
  W (I) = 0.0
20 Continue
  Do 30 N = 1, N Max
  Do 40 I = 1,9
  F = ((W(I+1) – 2.0 * W (I) + W(I – 1))/ (Dx ** 2)) + Q (I)
  WT (I) = ((1.0 – KS) * WT (I) + F * DT/ RHO)/ (1.0 + KS)
40 Continue
  Do 50 I = 1,9
  W (I) = W (I) + WT (I) * DT
50 Continue
C Boundary Conditions
  W (0) = 0.0
  W (10) = 0.0
C Check Instability
  NL = 0
  IF (W (5) . GT . 5.0) Then
  NL = 1
  GoTo 61
  Else
  Continue
  Endif
C Convergence Criterion
  LL = 0
  Do 31 I = 0,10

```
    IF (WT (I) . GT . 0.000001 . OR . WT (I) . LT . 0.000001) LL = LL + 1
31  Continue
    IF (LL . GT . 0) THEN
    GoTo 30
    Else
    GoTo 100
    Endif
30  Continue
100 Write (6, *) 'Number of iterations = ', NMAX
    Write (6, 8)
    Do 60 I = 0,10
    Write (6, 9) X (I), W (I)
60  Continue
61  IF (NL . EQ . 1) WRITE (6, *) 'Numerical instability is experienced'
    Stop
    End
```

The FORTAN program entitled Osama 2. FOR which is shown below is used to solve ordinary equation of example (2.1) using exact solution method.

Osama 2. FOR

```
C   This program solves an ordinary differential equation using exact solution
C   method
    Dimension W (0:10), X (0:10)
    Dx = 0.1
    PI = 3.1416
    Do 10 I = 0,10
    X (I) = DX * I
10  Continue
    Do 20 I = 1,9
    W (I) = sin ( PI * X (I) )
```

20 Continue

    Write (6, 8)

8   Format (1x, 'X', 8x, 'W')

    Do 30 I = 0,10

    Write (6, 9) X (I), W (I)

9   Format (1x, F4.2, 5x, F6.4)

30 Continue

    Stop

    End

2.2.2 Solution of Partial Differential Equations:

Example (2.2):

Using the dynamic relaxation (DR) method solve the following partial differential equation:

$$\frac{\partial^2 u}{\partial x^2} + \frac{\partial^2 u}{\partial y^2} + 2\pi^2 u = 0 \qquad (2.2)$$

Subject to boundary conditions: $u(x, 0) = 0, u(x, 0.5) = sin(\pi x), u(0, y) = 0, u(0.5, y) = sin(\pi y)$

The exact solution is as follows:

$$u = sin(\pi x) sin(\pi y)$$

The Computer output is listed in table (2.2) below. The first row of each set is the approximate solution whereas the second row is the exact value. The solution of this example was converged in 73 iterations.

Table (2.2) solution of example (2.2)

0.0000	0.0000	0.0000	0.0000	0.0000	0.0000
0.0000	0.0000	0.0000	0.0000	0.0000	0.0000
0.0000	0.0961	0.1826	0.2511	0.2948	0.3091
0.0000	0.0956	0.1818	0.2502	0.2941	0.3091
0.0000	0.1826	0.3472	0.4775	0.5606	0.5880
0.0000	0.1818	0.3457	0.4758	0.5593	0.5880
0.0000	0.2511	0.4775	0.6568	0.7712	0.8092
0.0000	0.2502	0.4758	0.6549	0.7698	0.8092

0.0000	0.2948	0.5606	0.7712	0.9060	0.9512
0.0000	0.2941	0.5593	0.7698	0.9048	0.9512
0.0000	0.3091	0.5880	0.8092	0.9512	1.0000
0.0000	0.3091	0.5880	0.8092	0.9512	1.0000

Equation (2.2) is written in finite differences form as follows:

$$f = \frac{1}{\Delta x^2}[u(i+1,j) - 2u(i,j) + u(i-1,j)]$$

$$+ \frac{1}{\Delta y^2}[u(i,j+1) - 2u(i,j) + u(i,j-1)]$$

$$+ 2\pi^2 u$$

The FORTRAN program entitled Osama 3. FOR which is shown below is used to solve ordinary differential equation of example (2.2) using DR method.

Osama 3. FOR

C This program solves a partial differential equation using DR method
   Real K, KS
   Dimension X (0:10), Y (0:10), U (0:10, 0:10), UT (0:10, 0:10)
   Open (unit = 5, File = 'Osama 3. Dat', status = 'old')
   Open (unit = 6, File = 'Osama 3. Out', status = 'unknown')
   Read (5, *) K, DT, RHO, NMax
8 Format (2x, 'U (I, J)')
9 Format (6 (2x, F 6.4))
   NL = 0
   DX = 0.1
   DY = 0.1
   PI = 3.1416
   KS = K * DT/ (2.0 * RHO)
   Do 10 I = 0,5
   X (I) = DX * I

10 Continue
    Do 11 J = 0,5
    Y (J) = DY * J
11 Continue
    Do 20 I = 0,5
    Do 20 J = 0,5
    U (I, J) = 0.0
    UT (I, J) = 0.0
20 Continue
    Do 30 N = 1, NMax
    Do 40 I = 1,4
    Do 40 J = 1,4
    F = (U (I+1, J) − 2.0 * U (I, J) + U (I-1, J))/ (DX ** 2) + (U (I, J+1)
    − 2.0 * U (I, J) + U (I, J-1))/ (DY ** 2) + 2.0 * PI ** 2 * U (I, J)
    UT (I, J) = ((1.0 − KS) * KS) * UT (I, J) + F * DT / RHO) / (1.0 + KS)
40 Continue
    Do 50 I = 1,4
    Do 50 J = 1,4
    U (I, J) = U (I, J) + UT (I, J) * DT
50 Continue
C Boundary conditions
    Do 41 I = 0,5
    U (I, 0) = 0.0
    U (I, 5) = sin (PI * X(I))
41 Continue
    Do 42 J = 0,5
    U (0, J) = 0.0
    U (5, J) = sin (PI * Y (J))
    42 Continue

C   Check Instability

　　IF (U (5, 5) . GT . 5.0) Then

　　NL = 1

　　GoTo 61

　　Else

　　Continue

　　Endif

C   Convergence Criterion

　　LL = 0

　　Do 31 I = 0,5

　　Do 31 J = 0,5

　　IF (UT (I, J) . LT . 0.000001) LL = LL + 1

31　Continue

　　If (LL . GT . 0) Then

　　Go To 30

　　Else

　　Go To 100

　　Endif

30　Continue

100 Write (6, *) 'Number of iterations = ', NMax

　　Write (6, 8)

　　Do 60 I = 0,5

　　Write (6, 9) (U (I, J), J = 0.5)

60　Continue

61　If (NL . EQ . 1) write (6, *) 'Numerical instability is experienced'

　　Stop

　　End

The FORTRAN program entitled Osama 4. FOR is illustrated below which is used to solve a partial differential equation of example (2.2) using the exact or analytical solution.

Osama 4. FOR

```
C   This program solves an ordinary differential equation
C   using exact solution method
    Dimension U (0:10, 0:10), X (0:10), Y (0:10)
    DX = 0.1
    DY = 0.1
    PI = 3.1416
    Do 10 I = 0,10
    Do 20 J = 0,10
    X (I) = DX * I
    Y (J) = DY * J
10  Continue
    Do 20 I = 1,9
    U (I, J) = sin ( PI * X (I) ) * sin ( PI * Y (J) )
20  Continue
    Do 30 I = 0,5
    WRITE (6, 9) ( U (I, J), J = 0,5 )
9   Format ( 6 (1 X, F 6.4) )
30  Continue
    Stop
    END
```

Example (2.3):

Solve the following system of partial differential equations over a square domain bounded by $0 \leq x \leq 1$ and $0 \leq y \leq 1$ using DR method.

$$\left. \begin{array}{l} \dfrac{\partial^2 u}{\partial x^2} + 2\dfrac{\partial^2 v}{\partial x \partial y} + \dfrac{\partial^2 u}{\partial y^2} - u + \dfrac{\partial w}{\partial x} = 0 \\ \dfrac{d^2 v}{dx^2} + 2\dfrac{\partial^2 u}{\partial x \partial y} + \dfrac{\partial^2 v}{\partial y^2} - v + \dfrac{\partial w}{\partial y} = 0 \\ \dfrac{\partial^2 w}{\partial x^2} + \dfrac{\partial u}{\partial x} + \dfrac{\partial v}{\partial y} + \dfrac{\partial^2 w}{\partial y^2} + q = 0 \end{array} \right\} \quad (2.3)$$

Where $q = \pi\left(\pi^2 + \dfrac{1}{2}\right) \sin(\pi x) \sin(\pi y)$

The boundary conditions are as given in the next figure (i.e. figure (2.3)).

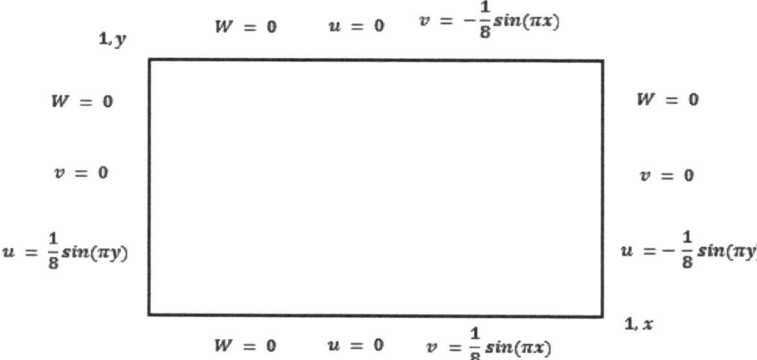

Figure (2.3) Boundary Conditions of Example (2.3)

Note that the differential equations are satisfied by the following solutions:

$$u = \frac{1}{8}\cos(\pi x) \sin(\pi y)$$

$$v = \frac{1}{8}\sin(\pi x) \cos(\pi y)$$

$$w = \frac{1 + 4\pi^2}{8\pi} \sin(\pi x) \sin(\pi y)$$

Write equation (2.3) in finite difference form:

$$f_1 = \frac{1}{\Delta x^2}[u(i+1,j) - 2u(i,j) + u(i-1,j)]$$

22

$$+ \frac{1}{4\Delta x \Delta y}[v(i+1,j+1) - v(i+1,j-1) - v(i-1,j+1) + v(i-1,j-1)]$$

$$+ \frac{1}{\Delta y^2}[u(i,j+1) - 2u(i,j) + u(i,j-1)] - u(i,j)$$

$$+ \frac{1}{2\Delta x}[w(i+1,j) - w(i-1,j)]$$

$$f_2 = \frac{1}{\Delta x^2}[v(i+1,j) - 2v(i,j) + v(i-1,j)]$$

$$+ \frac{1}{4\Delta x \Delta y}[u(i+1,j+1) - u(i+1,j-1) - u(i-1,j+1) + u(i-1,j-1)]$$

$$+ \frac{1}{\Delta y^2}[v(i,j+1) - 2v(i,j) + v(i,j-1)] - v(i,j)$$

$$+ \frac{1}{2\Delta y}[w(i,j+1) - w(i,j-1)]$$

$$f_3 = \frac{1}{\Delta x^2}[w(i+1,j) - 2w(i,j) + w(i-1,j)]$$

$$+ \frac{1}{2\Delta x}[u(i+1,j) - u(i-1,j)] + \frac{1}{2\Delta y}[v(i,j+1) - v(i,j-1)]$$

$$+ \frac{1}{\Delta y^2}[w(i,j+1) - 2w(i,j) + w(i,j-1)] + q(i,j)$$

Compute the velocities:

$$u_t(i,j)_n = \frac{1}{1+k_u^*(i,j)}\left\{[1-k_u^*(i,j)]\,u_t(i,j)_{n-\frac{1}{2}} + \left(\frac{f_1 \Delta t}{\rho_u(i,j)}\right)_{n+\frac{1}{2}}\right\}$$

$$v_t(i,j)_n = \frac{1}{1+k_v^*(i,j)}\left\{[1-k_v^*(i,j)]\,v_t(i,j)_{n-\frac{1}{2}} + \left(\frac{f_2 \Delta t}{\rho_v(i,j)}\right)_{n+\frac{1}{2}}\right\}$$

$$w_t(i,j)_n = \frac{1}{1+k_w^*(i,j)}\left\{[1-k_w^*(i,j)]\,w_t(i,j)_{n-\frac{1}{2}} + \left(\frac{f_3 \Delta t}{\rho_w(i,j)}\right)_{n+\frac{1}{2}}\right\}$$

Compute the values of $u(i,j)$, $v(i,j)$, and $w(i,j)$, from the following equations:

$$u(i,j)_{n+\frac{1}{2}} = u(i,j)_{n-\frac{1}{2}} + u_t(i,j)_n \Delta t$$

$$v(i,j)_{n+\frac{1}{2}} = v(i,j)_{n-\frac{1}{2}} + v_t(i,j)_n \Delta t$$

$$w(i,j)_{n+\frac{1}{2}} = w(i,j)_{n-\frac{1}{2}} + w_t(i,j)_n \Delta t$$

Apply the boundary conditions. If the domain is divided into 100 dements: 10 elements in the x – direction and 10 elements in the y – direction, then:

$$w(0,j) = w(10,j) = 0$$

$$u(0,j) = -u(10,j) = \frac{1}{8}\sin\pi(j\,\Delta y)$$

$$v(0,j) = v(10,j) = 0$$

$$w(i,0) = w(i,10) = 0$$

$$u(i,0) = u(i,10) = 0$$

$$v(i,0) = -v(i,10) = \frac{1}{8}\sin\pi(j\,\Delta x)$$

After each iteration check is made on the convergence and stability of the solution. The solution is considered converged when the velocities all over the domain is less than 10^{-6}. The criteria for instability is set by taking the bounds on w as: $0 \le w \le 5$. When solution has converged a further check is made for convergence to an invalid answer.

One must remember to exploit symmetry if it exists, a facility provided by the computer program listed next. In this example the solution can be obtained over one quarter of the domain using the following boundary conditions:

$$w(6,j) = w(4,j) = 0, v(6,j) = v(4,j) = 0, u(6,j) = -u(4,j) = 0$$

$$w(i,6) = w(i,4) = 0, v(i,6) = v(i,4) = 0, u(i,6) = u(i,4) = 0$$

The FORTRAN program entitled Osama 5. FOR is shown below which is used to solve a system of partial differential equations of example (2.3) using DR method.

Osama 5. FOR

C This program solves a system of partial differential equations
C using DR method
   Real K, KS
   Dimension X (0:10), Y (0:10), U (0:10, 0:10), V (0:10, 0:10), W (0:10, 0:10),
   UT (0:10, 0:10), VT (0:10, 0:10), WT (0:10, 0:10), Q (0:10, 0:10)
   Open (unit = 5, File = 'Osama 5. Dat', status = 'old')
   Open (unit = 6, File = 'Osama 5. Out', status = 'unknown')
   Read (5, *) K, DT, RHO, NMAX
8   Format (1X, 'U (I, J)')
13 Format (6 (1X, F 6.4))
9   Format (1X, 'V (I, J)')
14 Format (6 (1X, F 6.4))
12 Format (1X, 'W (I, J)')
15 Format (6 (1X, 6.4))
   NL = 0
   DX = 0.1
   DY = 0.1
   PI = 3.1416
   KS = K * DT/ (2.0 * RHO)
   Do 10 I = 0,10
   X (I) = DX * I
10 Continue
   Do 11 J = 0,10
   Y (J) = DY * J
11 Continue
   Do 20 I = 1,10
   Do 20 J = 1,10
   Q (I, J) = PI * ((PI ** 2) + 0.5) * sin (PI * X (I)) * sin (PI * Y (J))

20 Continue

    Do 30 I = 0,10

    Do 30 J = 0,10

    U (I, J) = 0.0

    UT (I, J) = 0.0

    V (I, J) = 0.0

    VT (I, J) = 0.0

    W (I, J) = 0.0

    WT (I, J) = 0.0

30 Continue

    Do 40 N = 1, NMAX

    Do 50 I = 1,9

    Do 50 J = 1,9

    F 1 = ((U (I+1, J) – 2.0 * U (I, J) + U (I-1, J)) / (DX ** 2)) +
        ((V (I+1, J+1) – V (I+1, J-1) –V (I-1, J+1) +V (I-1, J-1)) /
        (4.0 * DX * DY)) + ((U (I, J+1) – 2.0 * U (I, J) + U (I, J-1)) /
        (DY ** 2)) – U (I, J) + ((W (I+1, J) – W (I-1, J)) / (2.0 * DX))

    F 2 = ((V (I+1, J) – 2.0 * V (I, J) + V (I-1, J)) / (DX ** 2)) +
        ((U (I+1, J+1) – U (I+1, J-1) –U (I-1, J+1) +U (I-1, J-1)) /
        (4.0 * DX * DY)) + ((V (I, J+1) – 2.0 * V (I, J) + V (I, J-1)) /
        (DY ** 2)) – V (I, J) + ((W (I, J+1) – W (I, J-1)) / (2.0 * DY))

    F 3 = ((W (I+1, J) – 2.0 * W (I, J) + W (I-1, J)) / (DX ** 2)) +
        ((U (I+1, J) – U (I-1, J))/ (2.0 * DX)) + ((V (I, J+1) – V (I, J-1)) /
        (2.0 * DY)) + ((W (I, J+1) – 2.0 * W (I, J) + W (I, J-1)) /
        (DY ** 2)) + Q (I, J)

    UT (I, J) = ((1.0 – KS) * UT (I, J) + F1 * DT / RHO) / (1.0 + KS)

    VT (I, J) = ((1.0 – KS) * VT (I, J) + F2 * DT / RHO) / (1.0 + KS)

    WT (I, J) = ((1.0 – KS) * WT (I, J) + F3 * DT / RHO) / (1.0 + KS)

50 Continue

```
      Do 60 I = 1,9
      Do 60 J = 1,9
      U (I, J) = U (I, J) + UT (I, J) * DT
      V (I, J) = V (I, J) + VT (I, J) * DT
      W (I, J) = W (I, J) + WT (I, J) * DT
60    Continue
C     Boundary Conditions
      Do 41 J = 0,10
      W (0, J) = 0.0
      W (10, J) = 0.0
      U (0, J) = 0.125 * sin (PI * Y (J) )
      U (10, J) = -0.125 * sin (PI * Y (J) )
      V (0, J) = 0.0
      V (10, J) = 0.0
41    Continue
      Do 42 I = 0,10
      W (I, 0) = 0.0
      W (I, 10) = 0.0
      U (I, 0) = 0.0
      U (I, 10) = 0.0
      V (I, 0) = 0.125 * sin (PI * X (I) )
      V (I, 10) = -0.125 * sin (PI * X (J) )
42    Continue
C     Check Instability
      IF (W (5,5). GT. 5.0) Then
      NL = 1
      GoTo 61
      Else
      Continue
```

```
      Endif
C     Convergence Criterion
      LL = 0
      Do 31 I = 0,10
      Do 31 J = 0,10
      IF (UT (I, J). LT. 0.000001 . OR. VT (I, J). LT . 0.000001 . OR.
      WT (I, J) . LT. 0.000001)) LL = LL + 1
31    Continue
      IF (LL. GT. 0) Then
      GoTo 40
      Else
      GoTo 100
      Endif
40    Continue
100   Write (6, *) ' Number of iterations = ', NMAX
      Write (6, 8)
      Do 70 J = 0,5
      Write (6, 13) (U (I, J), I = 0,5)
70    Continue
      Write (6, 9)
      Do 80 J = 0,5
      Write (6, 14) (V (I, J), I = 0,5)
80    Continue
      Write (6, 12)
      Do 90 J = 0,5
      Write (6, 15) (W (I, J), I = 0.5)
90    Continue
61    IF (NL. EQ. 1) write (6, *) ' Numerical instability is experienced'
      Stop
      END
```

Osama 6. FOR

The FORTRAN program entitled Osama 6. FOR is shown below which is used to solve a system of partial differential equations of example (2.3) using exact solution.

```
C   This program solves a system of partial differential equations
C   using exact solution
    Dimension X (0:10), Y (0:10), U (0:10, 0:10), V (0:10, 0:10), W (0:10, 0:10)
7   Format (2X, 'U (I, J)')
8   Format (6 (2X, F 6.4) )
13  Format ( 2X, 'V (1, J)' )
9   Format ( 6 (2X, F 6.4) )
14  Format (2X, 'W (I, J)')
12  Format ( 6 (2X, F 6.4) )
    DX = 0.1
    DY = 0.1
    PI = 3.1416
    Do 10 I = 0,5
    X (I) = DX * I
10  Continue
    Do 11 J = 0.5
    Y (J) = DY * J
11  Continue
    Do 20 I = 1,5
    Do 20 J = 1,5
    U (I, J) = 0.125 * cos (PI * X(I) ) * sin (PI * Y(J) )
    V (I, J) = 0.125 * sin (PI * Y(I) ) * cos (PI * Y(J) )
    W(I, J) = ((1.0 + 4.0 * PI **2)/ (8.0 * PI)) * sin (PI * X(I) ) * sin (PI * Y (J))
20  Continue
    Write (6, 7)
```

　　　　Do 30 J = 0,5

　　　　Write (6, 8) (U (I, J), I = 0,5)

30　Continue

　　　　Write (6, 13)

　　　　Do 40 J = 0,5

　　　　Write (6, 9) (V (I, J), I = 0,5)

40　Continue

　　　　Write (6, 14)

　　　　Do 50 J = 0,5

　　　　Write (6, 12) (W (I, J), I = 0,5)

　　　　Stop

　　　　END

With my best wishes

Osama Mohammed Elmardi Suleiman

Mechanical Engineering Department

Faculty of Engineering & Technology

Nile Valley University

Atbara, Sudan

Bibliography:

[1] Rushton K.R., 'Large deflection of variable thickness plates', International Journal of Mech. Sciences, Vol. 10, (1968), PP. (723 – 735).

[2] Cassel A.C. and Hobbs R.E., 'Numerical Stability of Dynamic Relaxation Analysis of Nonlinear Structures', International Journal for Numerical Methods in Engineering, Vol. 35, No. 4, (1966), PP. (1407 – 1410).

[3] Day A.S., 'An Introduction to Dynamic Relaxation', The Engineer, Vol. 219, No. 5688, (1965), PP. (218 – 221).

[4] S.P. Timoshenko, 'History of Strength of Materials', McGraw – Hill, New York, (1953).

[5] J. R. H. Otter, 'Computations for prestressed concrete reactor pressure vessels using dynamic relaxation', Nuclear Structural Engineering 1, (1965), PP. (61 – 75).

[6] L.C. Zhang, 'Dynamic relaxation solution of elastic circular plates in large deflection under arbitrary axisymmetric loads', The third "5.4" Scientific Conference of Peking University, Peking University, Beijing, P.R. China, May (1987).

[7] S.W. Key et al., 'Dynamic relaxation applied to quasi – static, large defection, inelastic response of axisymmetric solids', In Nonlinear Finite Element Analysis in Structural Mechanics (Edited by W. Wunderlich et al.), Springer, New York, (1981), PP. (585 – 620).

[8] G.T. Lim and G.J. Torvey, 'On the elastic – plastic large deflection response of stocky annular steel plates', Computer and structure, 21, (1985), PP. (725 – 736).

[9] P.A. Frieze et al., 'Application of dynamic relaxation to the large deflection elasto – plastic analysis of plates', Computer and structure, 8, (1978), PP. (301 – 310).

[10] Aalami B., 'Large Deflection of Elastic Plates under Patch Loading', Journal of Structural Division, ASCE, Vol. 98, No. ST 11, (1972), PP. (2567 – 2586).

[11] Putcha N.S. and Reddy J.N., 'A refined Mixed Shear Flexible Finite Element for the Non – Linear Analysis of Laminated Plates', Computers and Structures, Vol. 22, No. 4, (1986), PP. (529 – 538).

[12] Turvey G. J. and Osman M. Y., 'Large Deflection Analysis of Orthotropic Mindlin Plates', Proceedings of the 12th Energy – Resources Tech. Conference and Exhibition, Houston, Texas, (1989), PP. (163 – 172).

[13] Turvey G.J. and Osman M.Y., 'Large Deflection effects in Antisymmet Cross – Ply Laminated Strips and plates', I.H. Marshall, Comosite Structures, Vol. 6, Paisley College, Scotland, Elsevier science publishers, (1991), PP. (397 – 413).

[14] Turvey G. J. and Osman M.Y., 'Elastic Large Deflection Analysis of Isotropic Rectangular Mindlin Plates', International Journal of Mech. Sciences, Vol. 22, (1990), PP. (1 – 14).

[15] M. Mehrabian, M.E. Golma Kani, 'Nonlinear Bending Analysis of Radial Stiffened Annular Laminated Plates with Dynamic Relaxation Method', Journal of Computers and Mathematics with Applications, Vol. 69, (2015), PP. (1272 – 1302).

[16] M. Huttner, J. Maca, P. Fajman, 'The Efficiency of Dynamic Relaxation Method in Static Analysis of Cable Structures', Journal of Advances in Engineering Software, (2015).

[17] Javad Alamatian, 'Displacement Based Methods for Calculating the Buckling Load and Tracing the Post Buckling Regions with Dynamic Relaxation method, Journal of Computers and Structures, Vols. 114 – 115, (2013), PP. (84 – 97).

[18] M. Rezaiee Pajand, S. R. Sarafrazi, H. Rezaiee, 'Efficiency of Dynamic Relaxation Method in Nonlinear Analysis of Truss and Frame Structures', Journal of Computers and Structures, Vol. 112 – 113, (2012), PP. (295 – 310).

[19] KYoung Soo Lee, Sang Eul Han, Taehyo Park, 'A simple Explicit Arc – Length Method using Dynamic Relaxation Method with Kinetic Damping', Journal of Computers and Structures, Vol. 89, Issues 1 – 2, (2011), PP. (216 – 233).

[20] J. Alamatian, 'A New Formulation of Fictitious Mass of the Dynamic Relaxation Method with Kinetic damping', Journal of Computers and Structures, Vol. 90 – 91, (2012), PP. (42 – 54).

[21] M. Rezaiee Pajand et al., 'A New Method of Fictitious Viscous Damping Determination for the Dynamic Relaxation Method', Journal of Computers and Structures, Vol. 89, Issues 9 – 10, (2011), PP. (783 – 794).

[22] B. Kilic, E. Madenci, 'An Adaptive Dynamic Relaxation Method for Quasi – Static Simulations using the Peridynamic Theory', Journal of Theoretical and Applied Fracture Mechanics, Vol. 53, Issue 3, (2010), PP. (194 – 204).

[23] C. Douthe, O. Baverel, 'Design of Nexorades or Reciprocal Frame Systems with the Dynamic Relaxation Method', Journal of Computers and Structures, Vol. 87, Issue 21 – 22, (2009), PP. (1296 – 1307).

[24] M. Mardi Osama, 'Verification of Dynamic Relaxation Method in the Analysis of Isotropic, Orthotropic and Laminated Plates using Large Deflection Theory', University of Shendi Journal, Volume 10, January (2011), PP. (31 – 52).

[25] Osama Mohammed Elmardi, 'Verification of Dynamic Relaxation (DR) Method in Isotroic, Orthotropic and Laminated Plates using Small Deflection Theory', International Journal of Advanced Science and Technology, Volume 72, Issue 4, (2014), PP. (37 – 48).

[26] Osama Mohammed Elmardi, 'Validation of Dynamic Relaxation (DR) Method in Rectangular Laminates using Large Deflection Theory', International Journal of Advanced Research in Computer Science and Software Engineering, Volume 5, Issue 9, September (2015), PP. (137 – 144).

[27] Osama Mohammed Elmardi, 'Nonlinear Analysis of Rectangular Laminated Plates Using Large Deflection Theory', International Journal of Emerging Technology and Research, Volume 2, Issue 5, September – October (2015), PP. (26 – 48).

[28] Javier Rodriguez Garcia, 'Numerical study of dynamic relaxation methods and contribution to the modeling inflatable life jackets', University of Brestagne Sud, (2011).

[29] L.C. Zhang and T.X. Yu, 'Modified Adaptive Dynamic Relaxation Method and its Application to Elastic – Plastic Bending and Wrinkling of Circular Plates', Computers and Structures, Volume 33, No. 2, (1989), PP. (609 – 614).

[30] L.C. Zhan and T. X. Yu, 'Application of mechanics of plasticity to the bending forming of beams and plates (in Chinese Language), Journal of Applied Science, 1, (1988), PP. (1 – 10).